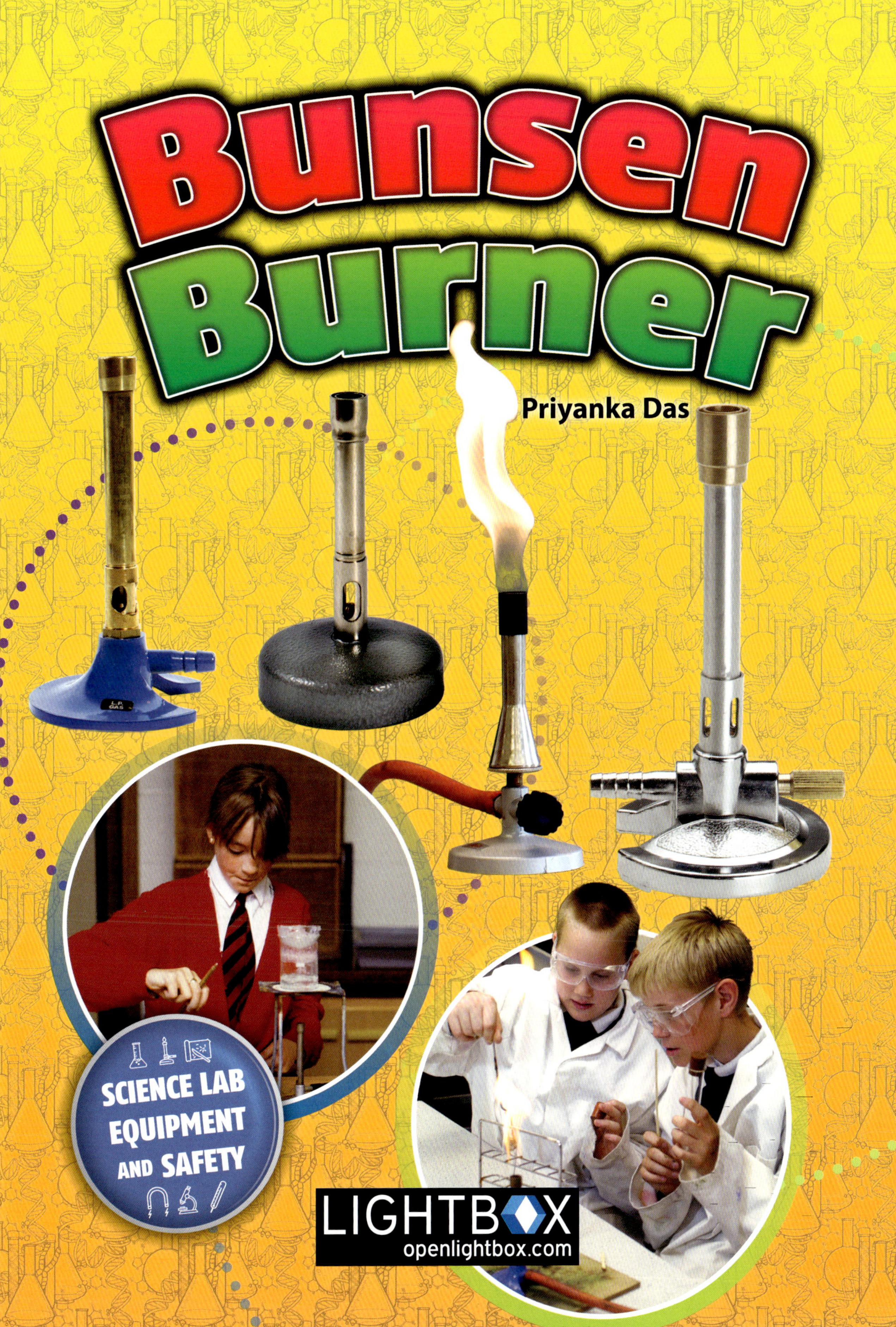
Bunsen Burner
Priyanka Das
L.P. GAS
SCIENCE LAB EQUIPMENT AND SAFETY
LIGHTBOX
openlightbox.com

Go to **www.openlightbox.com** and enter this book's unique code.

ACCESS CODE

LBXB8534

Lightbox is an all-inclusive digital solution for the teaching and learning of curriculum topics in an original, groundbreaking way. Lightbox is based on National Curriculum Standards.

LIGHTBOX SUPPLEMENTARY RESOURCES

SHARE
Share titles within your Learning Management System (LMS) or Library Circulation System

CURRICULUM
Find national and state curriculum correlations

CITATION
Create bibliographical references following the Chicago Manual of Style

STANDARD FEATURES OF LIGHTBOX

AUDIO High-quality narration using text-to-speech system

VIDEOS Embedded high-definition video clips

ACTIVITIES Printable PDFs that can be emailed and graded

WEBLINKS Curated links to external, child-safe resources

SLIDESHOWS
Pictorial overviews of key concepts

TRANSPARENCIES
Step-by-step layering of maps, diagrams, charts, and timelines

INTERACTIVE MAPS
Interactive maps and aerial satellite imagery

QUIZZES Ten multiple-choice questions that are automatically graded and emailed for teacher assessment

KEY WORDS
Matching key concepts to their definitions

Lightbox Grades 3–5 Subscription
ISBN 978-1-5105-5424-5

Access hundreds of Lightbox titles with our digital subscription. Sign up for a **FREE** subscription trial at **www.openlightbox.com/trial**

Bunsen Burner

CONTENTS

BURNING BRIGHTLY

Does your school have a science laboratory, or lab? The science lab is a place of discovery. It helps you understand how the things around you work. Many famous scientists have used science labs for **experiments**. Some of these experiments led to important **inventions**. Thomas Edison developed an electric light bulb after thousands of experiments in his lab. You can **conduct** experiments in a science lab, too.

Science labs have different **tools** that people can use. One useful tool is the Bunsen burner. Many experiments need heat, and the Bunsen burner is a heat source. It produces a hot flame. Using a Bunsen burner can be one of the most exciting parts of being in a science lab.

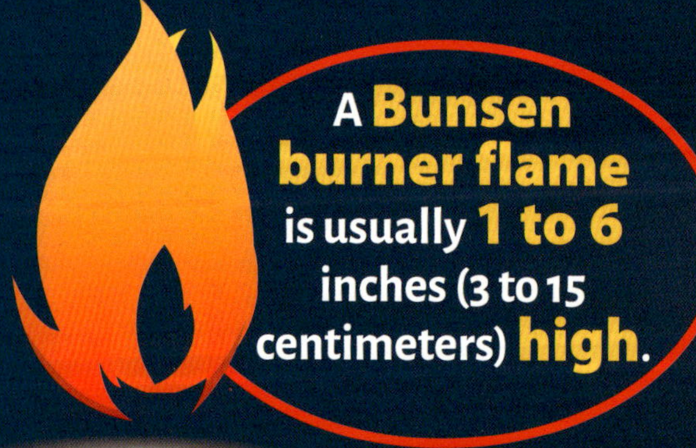

BUNSEN BURNER HISTORY

The Bunsen burner is named after Dr. Robert Bunsen. Bunsen was a German chemist and inventor who lived in the 1800s. He worked as a professor at a university. At the time, scientists were already using gas burners in labs. These burners had flickering flames that were difficult to control. Bunsen and Peter Desaga, a mechanic, improved the design of the gas burner. This improved burner became known as the Bunsen burner.

In addition to his career in chemistry, Bunsen had a lifelong interest in geology, which is the study of Earth's rocks and physical structure.

Today, the Bunsen burner is an important part of most science labs. Stable heat is often needed in experiments. The hot flame that the Bunsen burner produces is both steady and easy to control. People can change the flame's temperature. The Bunsen burner used in modern labs is not very different from the one Bunsen helped develop.

Robert Bunsen Timeline

1811
Robert Bunsen is born in Göttingen, Germany, in March.

1852
The University of Heidelberg hires Bunsen as a chemistry professor.

1855
Bunsen introduces the Bunsen burner.

2021
A medal named after Robert Bunsen is awarded to Professor Urs Schaltegge for his work in the field of geology.

PARTS OF A BUNSEN BURNER

A Bunsen burner is made almost entirely of metal. It consists of a metal tube attached to a flat metal base. The only exception is the rubber tubing. It connects the burner to the gas supply.

Gas valve
Controls the supply of gas

Rubber tubing
Carries gas to burner

Flame (outer cone)
Flame (inner cone)
Barrel
Metal tube where air and gas mix
Collar
Controls the supply of air
Air hole
Allows air into burner
Base

HOW IT WORKS

Before a Bunsen burner can be turned on, it needs to be connected to a gas supply. This provides fuel for the burner. The gas valve controls the amount of gas that enters the burner.

Gas mixes with air in the burner's barrel. This air enters the barrel through the air hole in the collar. The collar is a small disk. It can be moved to cover some or all of the air hole, controlling how much air enters the burner. Both gas and **oxygen** from air are needed to produce a flame. More gas means a larger flame, while more oxygen leads to a hotter flame.

The color of a flame changes with temperature. Blue flames are hotter than yellow flames.

When the collar covers the air hole, the burner produces a yellow flame. This is called a safety flame. It is the coolest flame and is very easy to see. When the air hole is open, more oxygen is available, and a steady blue flame is produced. Scientists use this flame to heat or burn materials in the lab. It has a blue inner cone and a violet-blue outer cone.

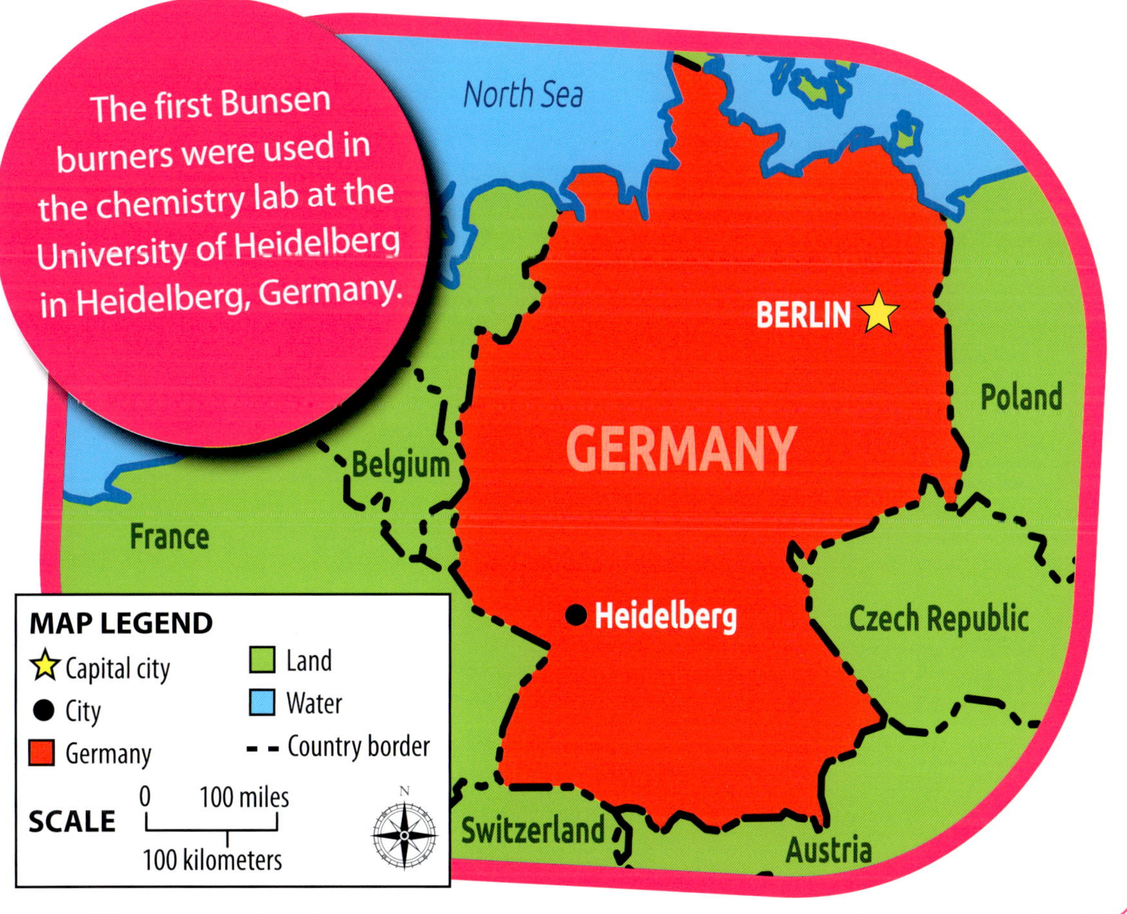

SAFETY IN THE LAB

Safety goggles protect people's eyes from hot liquids and other substances in beakers and flasks.

It is important to be careful in a science lab. Before using any tool, people must learn lab safety rules. The Bunsen burner is one of the most dangerous tools in the lab. If used incorrectly, it is a fire **hazard**. People can be burned by the flame. **Flammable** materials may catch fire.

Always wear safety goggles when using a Bunsen burner. Eye protection is often needed in the science lab, especially around an open flame. Tie back long hair and avoid loose clothing, too. These **precautions** will help prevent accidents.

Keep the area around the burner clear of anything that could catch fire. This includes paper, notebooks, and chemicals. Do not reach across the burner or lean over it. When the Bunsen burner is no longer being used, it can be switched off by turning the gas valve.

A Bunsen burner flame can reach about **2,700° Fahrenheit** (1,500° Celsius).

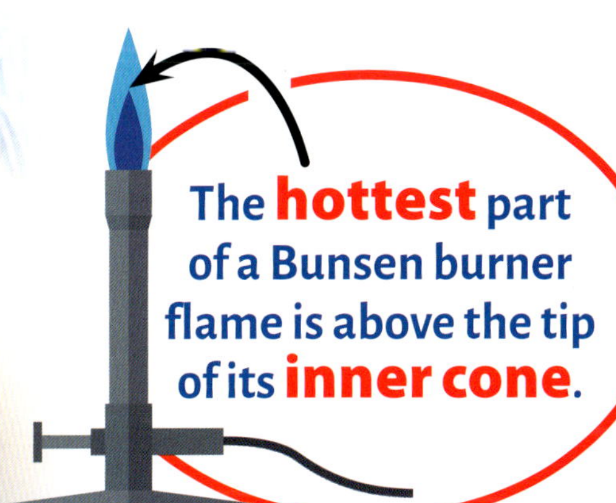

The **hottest** part of a Bunsen burner flame is above the tip of its **inner cone**.

THE SCIENTIFIC METHOD

The scientific method helps people understand how the world works. Scientists use it as a way to answer questions. The method involves testing a hypothesis with an experiment.

1. QUESTION

Ask a question about something that you have observed.

2. RESEARCH

Use the library and online research to find background information. Collect existing data that will help you answer your question.

3. HYPOTHESIS

Write down a **prediction** or a potential answer to the question. This should be an educated guess based on your research.

4. EXPERIMENT

Test your hypothesis by designing and performing an experiment.

5. OBSERVATION

Observe the results of your experiment. What happened?

6. ANALYSIS

Analyze the data you have collected. Determine the cause and effect of what you have observed. A cause is the reason that something happens, and an effect is what has happened.

7. CONCLUSION

Did the experiment support the hypothesis? Decide whether to accept or reject the hypothesis. Communicate your results to others.

EXPERIMENT

MYSTERY METAL

Lisa is a young scientist. Her teacher gives her a piece of metal and asks her to identify it. He tells her it could be copper, iron, or tin. After some research, Lisa decides the metal is copper. Can she prove it?

1. QUESTION What type of metal does Lisa have?

2. RESEARCH The piece of metal that Lisa was given is reddish-orange. It has a bright metallic shine, and it bends easily. Lisa looks up the properties of copper, iron, and tin. She learns what each metal looks like and how it reacts to fire. Based on its appearance, Lisa believes that the piece of metal is copper.

SAFETY FIRST!

Do not hold anything to the flame with your fingers. Always use a pair of tongs. After you are done, carefully put the tongs down. Make sure not to touch the hot end.

3. HYPOTHESIS

The mystery metal is copper.

4. EXPERIMENT

Lisa turns on the Bunsen burner. She uses a pair of **tongs** to lift the mystery metal to the flame.

5. OBSERVATION

The flame turns green!

6. ANALYSIS

Many metals can change the color of a flame. Copper is one of them. Lisa knows from her research that copper will turn a flame green. In the experiment, heating the mystery metal caused a green flame.

7. CONCLUSION

The experiment supported the hypothesis. The metal is copper.

Copper is a very useful metal. It is found in electrical wires, pipes, and some pots and pans.

EXPERIMENT

MELTING CHOCOLATE

Have you ever seen an egg fried on the stove? Heat changes a raw egg into a cooked egg. This change cannot be reversed. However, other changes are **reversible**. For example, an ice cube can melt back into water. Jamie wants to know what will happen if he heats a solid chocolate bar. Can this change be reversed?

1. QUESTION

Can heated chocolate be changed back into a chocolate bar?

2. RESEARCH

Jamie researches what happens when chocolate is heated. He finds out that heat makes chocolate turn into a liquid. This is called melting. Jamie also learns that melting can be reversed by freezing. He thinks that liquid chocolate can be made solid again.

SAFETY FIRST!

Never leave an open flame unattended. If the Bunsen burner is on, you must stay by it. If you need to step away, turn off the burner.

3. HYPOTHESIS Melted chocolate can be changed back into a solid bar.

4. EXPERIMENT Jamie puts a chocolate bar into a beaker. He places the beaker on a wire mat on top of a stand. Then, he lights the Bunsen burner under the stand. The chocolate slowly melts into a liquid. Jamie picks up the beaker with a pair of tongs. He pours the liquid chocolate into a small rectangular tray and puts the tray in a freezer.

5. OBSERVATION After about 20 minutes, the melted chocolate freezes in the tray. It turns into a solid bar again.

6. ANALYSIS Heating solid chocolate melts it into a liquid. Freezing liquid chocolate causes it to turn solid.

7. CONCLUSION The experiment supported the hypothesis. Melting chocolate is a reversible change.

Chocolate usually melts at temperatures between 86°F and 90°F (30°C and 32°C).

ACTIVITY

DESIGN A POSTER

Experiments follow steps. It is important that these steps are done in the correct order. Otherwise, the experiment will fail.

Scientists also follow steps to light a Bunsen burner. The different parts of the burner must be handled carefully. Design a poster that shows the steps people must follow in order to light a Bunsen burner.

STEPS

1. Using books and online research, find out how to light a Bunsen burner correctly.

2. Write the steps in the correct order on the poster and number them.

3. Illustrate each step.

HELPFUL HINT
Here's a hint to get you started! The first step is to check that all safety protocols are being followed. Turning off the gas by closing the gas valve is the last step.
Poster paper
Colored pencils or markers

QUIZ

1 What material are Bunsen burners mostly made of?

2 When was Dr. Robert Bunsen born?

3 In which country is the University of Heidelberg?

4 When is it safe to leave an open flame unattended?

5 What color will copper metal cause a flame to turn?

6 How is a hypothesis tested?

7 What is the hottest part of a Bunsen burner's flame?

8 Is frying an egg a reversible change?

9 What are the two parts of a blue Bunsen burner flame?

10 In which part of the Bunsen burner do gas and air mix?

Answers:
1. Metal **2.** 1811 **3.** Germany **4.** Never **5.** Green **6.** With an experiment **7.** The tip of its inner cone **8.** No **9.** Inner cone and outer cone **10.** The barrel

KEY WORDS

conduct: carry out

experiments: scientific procedures performed to discover something

flammable: easily set on fire

hazard: source of danger

illustrate: add pictures to something

inventions: useful devices or processes that have been created for the first time

oxygen: a colorless gas that is found in air and is necessary for fire to burn

precautions: actions taken to prevent something dangerous from happening

prediction: statement about what might happen in the future

reversible: capable of being returned to a previous state

tongs: tools that can be used to pick up and hold items

tools: instruments or devices that perform a particular function

INDEX

Published by Lightbox Learning
276 5th Avenue, Suite 704 #917
New York, NY 10001
Website: www.openlightbox.com

Library of Congress Cataloging-in-Publication Data

Names: Das, Priyanka, author.
Title: Bunsen burner / Priyanka Das.
Description: New York, NY : Smartbook Media Inc., [2022] | Series: Science lab equipment and safety | Includes index. | Audience: Grades 2-3
Identifiers: LCCN 2021032113 (print) | LCCN 2021032114 (ebook) | ISBN 9781510559059 (library binding) | ISBN 9781510559066
Subjects: LCSH: Science--Experiments--Safety measures--Juvenile literature. | Bunsen burner--Juvenile literature.
Classification: LCC Q182.3 .D37 2022 (print) | LCC Q182.3 (ebook) | DDC 502.8--dc23
LC record available at https://lccn.loc.gov/2021032113
LC ebook record available at https://lccn.loc.gov/2021032114

Printed in Guangzhou, China
1 2 3 4 5 6 7 8 9 0 25 24 23 22 21

092021
111020

Project Coordinator: Priyanka Das **Designer:** Ana María Vidal

Every reasonable effort has been made to trace ownership and to obtain permission to reprint copyright material. The publisher would be pleased to have any errors or omissions brought to its attention so that they may be corrected in subsequent printings. The publisher acknowledges Alamy, Getty Images, and Shutterstock as its primary image suppliers for this title.